Wit and Wisdom from the Construction Site

31 Life Lessons from the Builders of Society

Sean C. Jones

CASSL Books

Copyright © 2022 by Sean C. Jones. A CASSL Books Publication

All rights reserved.

No portion of this book may be reproduced without written permission from the publisher or author, except as permitted by U.S. copyright law.

The content contained within this book may not be reproduced, duplicated, or transmitted without direct written permission from the author or the publisher.

Under no circumstances will any blame or legal responsibility be held against the publisher, or author, for any damages, reparation, or monetary loss due to the information contained within this book, either directly or indirectly.

Legal Notice:

This book is copyright protected. It is only for personal use. You cannot amend, distribute, sell, use, quote, or paraphrase any part, or the content within this book, without the consent of the author or publisher.

Disclaimer Notice:

Please note the information contained within this document is for educational and entertainment purposes only. All effort has been executed to present accurate, up-to-date, reliable, and complete information. No warranties of any kind are declared or implied. Readers acknowledge the author is not engaged in rendering legal, financial, medical, or professional advice. The content from this book has been derived from various sources, including life experiences.

By reading this document, the reader agrees that under no circumstances is the author responsible for any direct or indirect losses incurred as a result of the use of the information contained within this document, including, but not limited to, errors, omissions, or inaccuracies.

Contents

Wit and Wisdom from the Construction Site V

1. Just Stop 1
2. It Gets Old 3
3. All You Can Eat 5
4. Idiot Proof? 7
5. Activity or Progress 10
6. Nine Women and a Baby 14
7. Sit on it! 16
8. Lunch Time 18
9. Fist Bump 20
10. Honey 24
11. Proud as a Peacock 26
12. Elephant in the Room 29
13. Experience 31
14. Help Wanted 34
15. Just A Minute 36

16.	Always Right	38
17.	Welds To Fix	40
18.	All the Hours	44
19.	I have to work late	46
20.	Don't be an Ass!	48
21.	Trifecta	50
22.	Play it Again, Sam	52
23.	Just Beat it	54
24.	I am all out of Names	56
25.	Chicken Salad	58
26.	Emojis don't count	60
27.	Do it for the Fishes	62
28.	What Time Is It?	64
29.	The Pen(cil) is Mightier than the Sword	66
30.	Designed to a "T"	68
31.	In the Navy	70

Wit and Wisdom from the Construction Site

After serving 25 years in the army I was not really looking forward to life as a civilian. The 9-5 desk job did not excite me in the least. Then I discovered the construction industry had a similar feel, in that there is a basic hierarchy, a meritocracy, and, most importantly, a sense of purpose and camaraderie and a little bit of hazardous work to round out the adrenaline junkie in many of us. The construction world had many of the qualities I loved about the army. To be sure, no one was shooting at me or things exploding around me, but that doesn't mean things were a cakewalk on the construction sites. Dozers and excavators were like working around tanks and other armored vehicles. Rigging for a lift with a crane was similar to sling loading cargo under helicopters. But the striking resemblance, the one that meant the most to me, was the people. Salt of the earth folks. Guys and gals that got up early every morning, often away from their families, and homes, to perform a job fraught with danger. All to provide a better life for their loved ones far away.

I have managed and supervised construction around the United States, Canada, and several locations in the Middle East. Regardless of the location, the superintendents, foremen, and workers were the same. They worked hard, were very knowledgeable, loved their families, and enjoyed their time off. The average construction worker may seem rough around the edges, but that is often a cover. They are intelligent; modern construction demands that they be. These men and women, like Soldiers and Marines around the globe, have a code of ethics they follow, a direct way of dealing with one another, and a language all their own. Much of the construction site language can be crude, but unlike what you see in television and movies, is not often vulgar. It is some of this language I wanted to capture.

The layout of this book is not linear. You do not have to read chapter one, then two, etc. Each chapter is a stand-alone piece. I had a wonderful time reliving these events in my mind as I wrote this. I hope you enjoy it as well.

Chapter One

Just Stop

Y**OU WILL FORGET THE times you stopped an unsafe act but never forget the time you let one slide and someone got hurt.**

I feel this statement is self-explanatory, but I am also surprised by the number of people I have seen walk right past a group of folks performing unsafe acts.

How often have I stopped a job and discussed the proper way to do a task or event? I have no earthly idea. And that is the point.

On the opposite side of this equation, I can honestly say I have not knowingly walked past something, let it slide, and seen someone get hurt. The key, however, is that one must identify and understand what can go wrong.

I feel such an occurrence would stick with me. The knowledge that someone lost a finger, broke a bone, or worse, and could have been prevented by me acting in their interest would haunt me for the rest of my life. Fortunately, this has not been my experience. Experience can be an unforgiving and brutal teacher. Please don't let it be yours! Show those on your site that you care. Let them see they mean more than hitting a schedule or coming in under budget.

Chapter Two

It Gets Old

WORKING SAFELY MAY GET **old, but so do those who practice it.**

This cannot be said enough. Daily, weekly, and monthly safety meetings can and often do get monotonous, with the ever-present possibility of complacency. They are, in the minds of many, a "check the block" activity to cover the backside of management and leadership.

But if one does care about the workforce on an individual level, imparting safety advice, tips, and tricks will help others to survive an unforgiving construction task. In other words, you get to live to fight another day. Take the time and invest in your workforce. Teach them so they can be proud of the work they do, not the work they did before they lost a finger. Later in life, they can tell stories of when dinosaurs roamed and they were Gods, and bounce grandchildren on their knees or, to show off what they built.

There is a similar phrase I can relate to that I learned in the army "Beware of old men in a profession where men die young." These older men did not get that way by breaking the rules or dumb luck. Karma is a bitch and will get those that skirt rules, and the odds are not in one's favor. Luck will run out.

Chapter Three

All You Can Eat

S AFETY IS FREE. **Use all that you want.**
I love this idiom. It cuts to the quick. There are many slogans about being your brother's keeper, looking after your fellow tradesman, etc. Still, ultimately, it comes down to one person deciding to perform an act that results in an incident. Maybe they hurt themselves or perhaps injured someone else when all they had to do was stop, think, then act. Is this safe? What is the worst that can happen? Is there someone I can talk to about this? All these acts of safety are free. There is a multitude of resources around that do not cost a dime.

It just requires a person to use those resources and as many as they want. It is not often you tell people to be selfish in a team environment. To me, safety is something I am selfish about. It is personal to me. Often times we talk about a potential incident's effect on others, be they spouses, kids, or coworkers, and leave out the effect on the one that got hurt. Most of us do not like pain. By definition it just sucks. So be selfish. Sit at the buffet of safety and grab another serving. It is free after all.

Chapter Four

Idiot Proof?

WE KEEP TRYING TO **idiot-proof this, but they keep giving us bigger idiots.**

They say one learns from experience, but the intelligent man learns from the experience of others. It is just human nature for some people to experience and learn things the hard way. The safety journey in the construction world attempts to teach others, so they do not have to learn the hard way.

In the world of safety, one takes many steps to mitigate hazardous work and make it as safe as possible. One of the steps is to verify the work has to be done. If there is no valid reason to do the work, then the problem is solved. Another step is to engineer the hazard out of the equation. If the work requires workers to be off the ground to do the work, as yourself "does it really?". Can we not provide integrated tie-off points into the structure? How about making sure there are suitable platforms, guards, and railings? Is there a way to perform the work on the ground?

Then some procedures can and should be implemented to do work as safely as possible. Work Permit processes exist to ensure work can be performed safely and that others are aware of the de-conflict potential interactions with other crews. Lock Out Tag Out (LOTO) procedures exist so that people working on motors, powered equipment, etc are free from worry they will get electrocuted or ground up in moving parts. Pre Startup Safety Reviews (PSSrs) walk teams through procedures to start new facilities, with an emphasis on all safety systems being operational and functioning.

And lastly, there is the use and application of the proper Personal Protective Equipment (PPE), the last line of defense.

In the ideal world, implementing the steps above should provide the greatest opportunity to execute the work and have everyone go home with all their digits and eyes intact. But.... we don't live in the ideal world all of the time. There is always someone that will find the chink in the armor. Someone who refuses to follow the procedure or does not wear all the protective equipment inevitably results in someone getting hurt or equipment getting damaged.

It is like a cat and mouse game, measure, countermeasure, measure, countermeasure. The only difference is the cat and mouse are playing the game knowingly. The safety journey is never complete and is constantly reacting and evolving. The "idiot" proofing, in the form of engineering controls, procedures, and PPE, often gets one-upped by the bigger "idiots."

Chapter Five

Activity or Progress

DON'T CONFUSE ACTIVITY WITH **progress**

This is similar to a quotation by Mary O'Connor, who famously said, "It is not how busy you are, but why you are busy — the bee is praised, the mosquito is swatted."

I had a project manager that was notorious for cutting through all the B.S. He asked very tough questions. Questions that he needed answers to for his upline reporting. He could be brutal if you didn't know the answers – a base expectation for your job, but what was worse is making up answers or tap dancing around the question. That would usually get you in the penalty box. One in which you got lectured at a very high volume for what seemed like hours Everyone on the team experienced this phenomenon at least once in their time on the project.

One instance that was emblazoned into my brain occurred when our project Quality Assurance (QA) manager tried to tap dance around the question. This resulted in, I kid you not, a one-hour tirade by the PM. What made this one most memorable though was the location of the interaction and the location of the project. The project was located in southern Iraq. A very inhospitable place in terms of weather. The people were amazingly good-hearted and hard-working, but that weather was not enjoyable. On this particular morning, the temperature was a mild 104F/40C. Mild by summer conditions which regularly went above 120F/50C during the summer. The exchange and I say this generously as it really was one way, took place on the main walking path to and from our offices. Everyone got to see this interaction at least once, some even tried to intervene politely with a new topic that needed the PM's attention. He was not to be distracted from his target. The PM wanted to make sure his intentions and expectations were clearly understood- and I think it was a little bit of a show to ensure EVERYONE on the team got the message After the QA manager was released, many people went by to offer their support to him in the form of cold drinks. He must have sweated out a gallon of liquid during that hour – heat exhaustion was a real possibility.

By this point, I am sure you are dying to know what was so important as to nearly dehydrate a man and take on a heat-related safety event. To be clear, this is more of an exaggeration, neither of them was in any danger of a heat injury.

It all stemmed from a question as to why the welding repair rate was so high. To be fair, it is more of a production question than a quality assurance question, but it was just his turn. The QA manager tried to explain that the contractor had plenty of personnel on site, and the welding quality control representatives were extremely active in ensuring the proper procedures were utilized, But simply having the staff present was not good enough. The point emphasized by the PM was that being busy, or active is great, but it doesn't put wood on the proverbial woodpile. One must be productive, not just active. The honey bee and the beaver are very productive. They are building something for the benefit of the hive or the den respectively. But the mosquito, the bane of the world, is merely active. They are only in it for themselves. Their only value is as food for the birds.

I dare say that almost everyone, at one point in their life, was guilty of "looking busy" when the teacher, boss, or even a spouse walked by. Being busy without purpose isn't good enough in the real world. So in the realm of active versus productive, do not be the mosquito, be the bee or beaver.

Chapter Six

Nine Women and a Baby

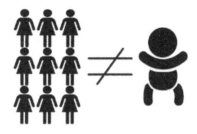

ONE WOMAN CAN HAVE a baby in nine months, but nine women can't have a baby in one month.

As a young construction lead on a capital project for the oil and gas sector, a very crusty construction manager from our contractor staff and I were having a discussion. The "discussion" centered around the lack of progress on the job. The contractor was weeks behind schedule.

When one looked around the job site, it was easy to ascertain why- there were hardly any workers on site. I made that simple observation to contractor CM and commented on how it is not hard to see why progress isn't being achieved. "You need more people out here. There is no way you can prove to me that you can catch up without staffing up."

His reply was simple, "Sean, one woman can have a baby in nine months, but nine women can't have a baby in one month. Throwing people at the problem is not always the solution."

His comment was not without merit. More people, in extreme quantities, lower productivity due to work face density being too high. In other words, there are more people than space to work efficiently. Additional concerns also poke their head up when you have more people or even additional shifts. The additional staff will require more significant resources to get the work done.... materials, tools, and support structures, to name a few.

From this interaction, I learned that more bodies are AN answer, but it is not the silver bullet most people assume. Those folks need to be supported, and ultimately, this is where the issue lies; the inability of the contractor to provide sufficient material to keep the work faces open and progressing. It is a vivid analog and one lesson I will never forget.

Chapter Seven

Sit on it!

WORKERS CANNOT JUST SIT **on their hands if you provide materials, drawings, and tools. They won't be stopped from putting things together.**

The lack of productivity was true, and the contractor was missing something; it wasn't workers. They did not have the materials to sustain activities. All the previous discussions about staffing would not have solved the clawing back-of-schedule issue. The contractor had more significant issues they were trying to keep from the owner team.

One of the sharpest construction hands I have ever worked with said this to me one day, shortly after the conversation involving our contractor having to staff up (see the previous chapter). He attributed his wise words to his father. Simply put, if everything is available, it is just human nature to start to put it together. There is no way the company was telling their people to slow production – it sounds crazy, but this was a theory being thrown around by senior management. Well, they may have been told to sit on their hands, but they weren't going to do that.

Construction folks love to tinker. They enjoy the hard work and satisfaction of completing complex tasks. It was killing them, morale-wise, to not be able to do what they are paid to do. Money for nothing sounds good, but it doesn't bring a sense of self-worth or accomplishment.

Chapter Eight

Lunch Time

G**O ON AND EAT it; it ain't but a little bit.**
One of the great traditions we had in the military units I was in was the monthly roasting of leadership for idiotic, stupid, or just plain messed-up actions during the previous month. The rules were simple, nominate someone for a said offense, tell the story (complete with embellishments- only 10% needed to be true), and at the end, we vote on who won the award for that month.

Some paratrooper units had the "Failure to Hook up Award" in the sense that you must hook up your parachute before you jump out of a perfectly good aircraft. One of my units had a Kodiak grizzly bear as its mascot.

Our award was the "Bear Ass Award," a little play on words.

The final rule was simple. No quibbling – the person nominated could not defend their actions; they had to sit there and take it – grin and bear it (so to speak). Thankfully, I never excelled to those heights and was never nominated.

Now, people did stupid shit all the time on the construction site. In much the same way as the military example, one would get roasted by peers and management for those actions, just not in an organized formal sort of way. Unlike the military rules of etiquette, the folks on the construction site always seemed to want to defend their actions, which was the equivalent of jumping into the hole they were in and helping to dig it deeper. Fighting back just invited others to pile on with more vigor, and some even brought up long-lost actions of the past just to show a trend.

When all this happened, one of my supers would eventually pipe up with his slow Alabama drawl and tell the hapless defender, "Go on and eat it; ain't but a little bit." He was right, eat a little of that humble pie, and the jackals and vultures soon lost interest. It is not a lot of fun trying to fight someone who isn't fighting back and has accepted their fate amongst their peers.

Chapter Nine

Fist Bump

THEY SAY A MAN'S **fist is as big as his heart. See what you just broke.**

This is a day I will never forget. I am all about the business at hand when on the job (See Chapter 28). We can handle niceties later. Little did I know, this was not to be one of those days.

So, there I was, on a scorching and humid day on a US Gulf Coast job site. I was making my rounds looking at progress, talking to the field craft, and checking to see if all were working safely, materials were on site and general welfare of the craft. I spotted a group of individuals performing what I considered a potentially unsafe act. I approached them and had a quick discussion. I do not recall the exact situation, and for the sake of this story, it doesn't matter. I knew my next stop was to see that company's safety representative and encourage him to look and lend expertise and support to the crew.

I spotted the safety man nearby; he was not hard to find. He was 6'3" tall and 295 pounds (2950 pounds if he had to come down on you, as he often exclaimed). As you might suspect, he went by the nickname "Big."

I approached Big and, as was my nature, jumped right into business, explaining what I had witnessed and was going to recommend he take a look and help out. I didn't get that far. Big interceded, with his booming voice laced with sarcasm, "Good morning, Sean. How are you doing this morning, Sean?"

He must have seen the puzzled look on my face and continued,

"It's always right to the problem with you; You never start with 'How's it going Big? How's the world treating you, Big?"

"You know, they say a man's fist is as big as his heart." He then shows me his fist – the size of a grizzly bear. "See what you just broke?"

At first, I thought he was being both overly sensitive and dramatic, but then I thought a little harder and apologized. I was trying to be efficient but also leaving out the human touch. The work crews were not in imminent danger. I did not need him to drop everything, run over, and save the day. I did have time to exchange pleasantries and continue developing the friendship and working relationship.

I had a similar occurrence working in Saudi Arabia. An Egyptian contractor PM came into my office and wanted to discuss an invoice he had sent me the previous day. I looked at him, and before exchanging words, I immediately started to tell him what was wrong with the invoice. He looked at me dejectedly. He took the invoice from my hands and walked out. His deputy, an American, dropped off the corrected invoice the next day. I did not see Mohammed for another 45 days. He was a crucial player in the work I needed to accomplish. Instead, he sent his deputy or senior superintendent from that day on.

On good terms with Mohammed, my office mate told me a couple of days later that Mohammed was distraught with me and felt like I did not treat him as a man. I later apologized, and we worked well together for the next ten months.

I learned from Big and Mohammed earlier (which didn't stick) that the construction business is about people first and foremost. The concrete and steel, the dozers and dumps, the cranes and excavators do not care how we do things. But the managers, the operators, the craftsmen, and the laborers all sure do.

While I expect people to get to the point with me, that is how I operate. Other personalities or cultural norms require or dictate a more personal approach. I had failed on both occasions.

To the best of my knowledge, those were the last times I allowed that to happen. Someone reading this may tell me later I did as much to them. And to them, I offer my apologies.

Chapter Ten

Honey

IF YOU TAKE CARE of the bees, the bees will provide you with honey.

I learned this from a Russian Electrical Superintendent on a job in Canada. He had the sharpest, hardest working crews on the job site. Their artistry was impeccable. Their safety statistics stood out as the best. The quality control and assurance folks often used their work as the standard others should strive to achieve.

One day, I asked Igor, the Superintendent, how he does it and how he gets the results. Even the next-best crews were not in his league. He replied, "In my country, (Russia) we have a saying. If you take care of the bees, the bees will provide you with honey." It is a beautiful metaphor for such a basic tenet of leadership. The US Horse Cavalry used to say "Feed your horses, feed your men, then feed yourself". Take care of your men; they will take care of you. His crew members were loyal to a man. They loved working for Igor. They understood, come hell or high water, Igor had their backs. Thus, they could focus on the tasks, let Igor drown out the background noise, and run interference when necessary.

Igor knew he could not be successful without the efforts of his crews. He was successful because he made them successful.

Chapter Eleven

Proud as a Peacock

PERSONALITIES ON A CONSTRUCTION site can often be larger than life. The job sites are filled with individuals that strive for excellence and do not like to be wrong. One of the challenges of leading such hard chargers is being able to focus their energies on areas that need their expertise. Often times egos can get in the way.

Sitting someone down, who has many more years of experience over you is a common occurrence in the junior ranks of the military officer corps. Young lieutenants straight out of school, while technically higher in rank, must lead the crusty Sergeant with years of combat experience. For the young Lieutenant, tact and a good working relationship provide a valuable source of knowledge and go a long way in accomplishing missions.

I found myself in similar situations in the construction industry. I vividly recall one incident in my office where the "crusty" superintendent was telling this young construction lead how it is going to be, He started off on a rambling story about watching a sports show on TV the previous night. The show featured the amazing Duke University basketball coach, Mike Krzyzewski. He talked about what a phenomenal leader Coach K is. How he gets in the trenches, set high standards, and expects nothing less. He then transitions into how he just found out Coach K was a graduate of The United States Military Academy at West Point. Large heapings of praise followed about the service academy and the leaders it has produced for this nation. He then pauses, looks me right in the eye, and says, "You graduated from West Point didn't you?"

I replied in the affirmative, and if you listen to him. he claims I sat up straight and puffed out my chest. He then simply said, "They must really be disappointed in you!"

Go on and eat it; ain't but a little bit.

It was all in good fun, but he has never let me forget that moment over the past 15 years. Amazingly, I call him a dear friend to this day in spite of his lack of understanding.

A year or two later in the project (it was a four-year build), I had him in my office and was trying to focus his energy on an area of concern on the site. After we talked it through, he understood what I needed him to do, and something was said about finally understanding leadership.

This is where I got to zing him back for the Coach K thing. I told him, that my job is very simple. I have a stable of peacocks, all with differing levels of plumage. It is my job to talk to those peacocks, smooth out their plumage, and send them on their way. Ending the conversation with some of my peacocks that need more smoothing out than others. He smiled and left my office.

It was a couple of days later when he told me he had been thinking about the peacock plumage metaphor a lot since I talked to him. I won't say his exact words, but they were something like, "You trucker!" The mind bomb was planted, and the time delay went off. It was my turn to smile.

To this day, when I see a peacock painting, Christmas tree ornament, or any manner of tchotchke I will send him a photo via text. It doesn't take him long to ask if I am allowed to wear my West Point ring yet.

The lesson here is to know your peacocks. Let them vent and get things off their chest. When that is done, calmly smooth out their ruffled feathers, give clear direction, and let those peacocks strut their stuff. They have earned their plumage.

Chapter Twelve

Elephant in the Room

WHEN ELEPHANTS FIGHT, DON'T **be the grass – attributed to a Nigerian saying.**

This little gem was a favorite of one of my Project Managers – it was an eloquent way for him to remind others to "butt out." Relatively simple, if two bosses (the elephants) are arguing a point –er, ah, discussing, I mean to say, there is no need for you (the grass) to jump in and get trampled. I have often seen young engineers trying to insert themselves to support one "elephant," only to have it backfire and get crushed by the other "elephant", unless they ask for your input. Let the big boys play it out, and then you will know what to do.

It is such a vivid picture and strong meaning that I have taught it to my twin daughters, who have adopted it when they argue with others. When one of them begins to defend their position to me or Mom, the other will often chime in for no reason other than to try and sway the argument in their favor. I was never prouder than when Twin A said to Twin B, "You're being the grass!"

Chapter Thirteen

Experience

*Knowledge is knowing a tomato is a fruit.
Experience is not serving it in a fruit bowl.*

THE ONLY WAY TO **get a foreman with 20 years of experience is to get a foreman who has worked for 20 years.**

Currently, there is a shortage of skilled craftsmen in the construction industry. Carpenters, welders, plumbers, HVAC technicians, and electricians, to name a few. Years of proselytizing college as the only reputable way to make a living have taken their toll. Skilled trades, a noble profession, have taken a big hit with the push that college is the only way. The belief that the blue-collar workforce is not a good way to make a living has been rampant. As a result, thousands of well-paying jobs have gone unfilled. The shortage means that older generations are working longer to keep projects moving forward. With fewer people entering the industry, the ability to pass on their body of knowledge diminishes. In the current and foreseeable future, skilled trades will become increasingly critical and most likely will become highly compensated.

In the meantime, knowledge transfer from the current craftsman to the future is essential. The industry has been trying to bring new hires quickly up the learning curve. One of the companies where I worked had summer interns interview and record conversations with the Subject Matter Experts (SMEs) in various fields. The result of this was two-fold. One, the intern got to sit face to face with a leading expert and soak in the fountain of knowledge firsthand, but this only benefited the intern. Two, the recorded interviews could be posted on the company server for current and future employees to glean knowledge from them; the intent, from the outside, looking in, would seem to be getting the one or two-year construction professional up to a level of a five or six-year person as quickly as possible, Shaving off years of development through the use of technology and productivity enhancements.

While this is a noble cause and can expedite the development levels much faster in certain areas, it is not a silver bullet. We learn through experience – ours or someone else's. It is not physically possible to take 20, 30, or 40 years of experience and compress it into eight one-hour podcasts. There are way too many experiences that the interviewee has forgotten that account for their depth of knowledge. You can only get twenty years of experience by investing 20 years into the field. Knowledge is not experience.

Chapter Fourteen

Help Wanted

Construction is an odd business. It is one in which the goal is to work yourself out of a job. You have rated a success, the faster you can become unemployed.

    A steady well paying job goes a long way to providing a stable life for a person and their family. Many people look for a job that has a defined career progression and long-term employment. Job stability matters to most people.

Now, imagine applying for a job knowing full well, that it only lasts for a year, six months, or some other defined period of time. Imagine further that the timeline is not guaranteed because your new employer wants to eradicate your job much quicker than the stated term. This is the world of construction. Some projects are very large, taking on the order of four to six years to complete. Hooking one of those jobs does provide one with a greater level of predictability in life. The three-month job however provides for a certain level of discomfort knowing that almost as soon as you get to know the job site, it's time to shut down the job and look for the next.

Time is money – rarely do owners want to spend more money than they have to. Slow and inefficient constructors do not get follow-on jobs. From a contractor's perspective, generally speaking, not having to carry indirect costs such as work trailers, lunch facilities, water, fuel etc provide a significant saving.

If you want to be considered one of the best in your field, you land the job and work to be unemployed as fast as possible. This gets you the next job, and the cycle continues.

Chapter Fifteen

Just A Minute

IF YOU WAIT UNTIL **the last minute, it only takes a minute.**

Anyone that ever crammed for a test in school or waited until the last possible moment to start their term paper knows the truth in this statement. Why do today what you can put off to tomorrow is one philosophy in execution? It is probably not the best choice in overall execution strategy, but a philosophy nonetheless. But as with all things, competing demands take our focus away from some things. That is usually where this comes into play.

Boss: "Don't forget you have the weekly safety topic for the meeting tomorrow. All set?"

You: "Nah, I'll get it done his afternoon."

Boss: "Well, don't wait until the last minute. It needs to be impactful."

You: "But Boss, if I wait until the last minute, it only takes a minute."

Many term papers, monthly reports, and other presentations throughout history have been put together in just this manner. Imagine if we put time into some of these ideas. We might make a difference. Until then, we only need a minute.

Chapter Sixteen

Always Right

THE ONLY SCHEDULE THAT **is accurate is the as-built schedule at the completion**

Even a broken clock is correct twice a day the old adage goes– of course, it has to be a broken analog clock. The project schedule is much like a clock. It is designed to tell you where you are at any given time.

No offense meant to schedulers and planners, as they put a lot of time and effort into producing a schedule, with logically linked predecessors and successors, meticulously researched durations for various construction activities, and intricate flowing Gannt charts, which look amazing but do not actually predict or forecast with any extreme certainty the conclusion of the job. Scheduler tricks would include terms like a P50 schedule – meaning we had a 50% chance of achieving that schedule end date. Then there is all the hidden time project managers keep in their back pockets. Contingency, schedule reserve, and so on. There is actually a movement now called, Scrum and Agile, two different schools of thought, which do away with the flowing Gannt charts and look at discrete blocks of work that can be done, with a claim of twice the work in half the time (J.V. Sutherland, 2015).

With all this said, Schedules are useful in setting up the expectation of a completion date. The worst way to try and execute a construction project is the fly-by-the-seat-of-your-pants, It will be what it will be attitude. The whole purpose of the schedule is to have a plan from which to deviate. Construction workarounds are a very frequent occurrence on job sites. As we used to say in the military, "the enemy gets a say in your plan." In construction, the enemy wears many hats – weather, labor, cost, and supply chain disruptions to name only a few. There is no way to predict what is going to cause issues in an hour, a day or a month from now.

While a schedule is a necessary tool for planning, the only schedule that is ever accurate, is the schedule of activities that have been completed – the actuals or as-builts – provided they are recorded when they occurred.

Chapter Seventeen

Welds To Fix

WELDS TO FIX!
No shit, so there we were, up to our knees in hand grenade pins, with nothing left to shoot but an azimuth. One of the classic beginnings of any war story. Here goes one from a time long ago.

Hurricane Ike struck the US Gulf Coast in September of 2008. It was a devastating hurricane that took many almost 200 lives and inflicted over $30 billion in damages to Haiti, Cuba, Louisiana, Texas, and even had effects on Ontario, Canada and, Iceland. It was a monster. (Wikipedia Contributors), My job site was no different, while we did not lose any human lives there were plenty of alligators that didn't make it, and the damage was massive. The storm surge that came ashore left standing water of 6 feet on the site. The effects of the storm surge coming in picked up and moved our job site trailers and vehicles. We found some of our process pipes miles away from the site for example. The pipe end caps kept the water out and made nice 40' long floating torpedos.

We spent almost 14 months digging out, cutting out, and moving out damaged materials, equipment and muck. Our storage tanks were massive at almost 100 yards in diameter and 200' tall, the beasts were encased in concrete. We thought for sure nothing would damage them, and structurally we were correct. What we did consider was the fact that the tanks had temporary construction openings which allowed water to flow into them and inundate the insulation inside, as well as start to have corrosive effects on the steel walls and floor plates. We had a long rehabilitation ahead of us. In the meantime, the management for the owner team and contractor team got into the who pays for what and how much is it going to cost battle. The sums of money were not insignificant, and time was exorbitant. But bad blood had been spilled. It got to the point where the contractor kept just enough folks on site to keep work progressing so as not to default on the contract. Sadly. this meant many craftsmen were now out of a job or moved on to other jobs. As the owner team, we had gone so far as to start looking at replacing the contractor and executing the work ourselves. It was not a fun time.

As if all this wasn't enough, we also hosted students from a construction management course the company conducted in Houston about once every 6 weeks so they could come out, see the construction sites and see firsthand what they were learning over PowerPoint. Once such a group I was hosting came on site, and I took them inside one of the tanks. These things are impressive and were usually the highlight of the tour. As I walked in, I spotted three gia-normous letters painted on the inside shell of the tank by an obviously disgruntled tradesman. In letters literally 8 feet tall were the initials "WTF". The old Whiskey Tango Foxtrot. If you are unfamiliar with the letters WTF, it stands for What The F&@k. I steered the tour group to the other side, but it was too late. An eagle-eyed student spotted the letter, and honestly, it wasn't hard to see them. I introduced the superintendent of tank construction, he proceeded to tell the class about the construction methods, materials, and how we have gone about the repairs. When he finished, he asked if there were any questions.

"What's that WTF on the wall mean?" asked the student.

without missing a beat, the super replied, "Welds To Fix. We had a lot of bad welds in that area that did not pass X-ray inspections. We have to cut out those panels and re-weld them."

Amazingly, the class bought the explanation and we concluded the tour.

I was never more impressed with that super than that moment. What an absolutely magnificent redirection and explanation he pulled out of his...

From that day forward, and even to this day, those of us on the project will still talk to each other and when something unbelievable happens, we look at each other and simply exclaim "Weld To Fix, man. Welds To Fix"

Chapter Eighteen

All the Hours

There are 24 hours in a day, and then we have the night shift!

Oh, the joys of working in multiple time zones! While working in the middle east, the days were long, hot, and dusty. We pushed hard to get the work completed during the day. With the heat, one hour felt like two. Those 12-hour days really took a toll and probably reduced our life span- kind of like dog years!

When the "workday" was almost complete, the engineering and design office, 9 hours behind our time zone, would start coming to work. Conference calls and emails would fly at the speed of light around the globe.

This was what we called the night shift. When you do the math, 12-hour days with the multiplier of 2 go us working a full 24 hours. Then the 8-hour night shift. For weeks on end, we put in 32-hour "days," 7 days a week. It wasn't literally 32 hours, obviously, but the toll on the body and mind certainly made one believe it so.

Chapter Nineteen

I have to work late

WORK UNTIL YOU DIE…THEN **stay a little longer – it's worth time and a half!**

I heard this recently when a Foreman was talking to some of his younger workers. The crews were asked to stay late and work some overtime to finish a task. It was tough dirty work, and most of them just wanted to shower and put on some clean clothes. The Foreman just brushed it off and told his crew to get to work. One of them responded with a smart-ass comment about being old and still working a back-breaking job and not wanting to be like him working himself to death. The foreman just smiled and replied, "I'm going to work until I die, then stay a little longer for the time and a half." That is dedication.

Chapter Twenty

Don't be an Ass!

WE WILL WORK THEM **like rented mules.** In an earlier chapter, we discussed a project that was behind schedule and various means to bring it back in line with the project schedule. After telling me that adding additional people to the site would not increase productivity, the CM told me we would extend the work hours and work through the next few weekends. I asked him how this would improve productivity when adding additional manpower will not. He went on to explain the support infrastructure was already in place. He said, and these were his exact words, "We'll just work them like rented mules."

Anyone who has done any extended-hour, extended-duration jobs knows that productivity decreases the longer the duration of hours. Construction Industry Institute and RS Means Company have studied and *documented the effects of extra hours and weeks. The reduced productivity is not insignificant. Going from a standard 40-hour work week(5 days x 8 hours) to a 50-hour week (5 days x 10 hours), the progress goes from the baseline of 100% to 85%. Increasing to 6 days x 10 hours shows reduced productivity to 80%.*[1]

This takes into account the lower production rate. It does not mention the increased occurrence of workplace incidents and errors in quality craftsmanship.

Don't be this guy! He was trying to impress me, show me how tough he was – it is not tough, it is abusing your people. Make sure to read the Honey chapter – take care of your people, and they will get it done. Going into the discussion with the attitude the workers are animals, and you can treat them like a rental (and we all know how rental cars are treated) will not get the job accomplished.

---

1. R.S. Means Company, 2014 pg 672.

Chapter Twenty-One

Trifecta

FAST, CHEAP, AND GOOD. **You can have two but cannot have all three.**

The three-legged stool, the holy trinity of construction: schedule, cost, and quality. Every project aims to achieve a project completed ahead of schedule, under budget, and with an excellent quality of craftsmanship. The problem is those three concepts have a built-in tension...by design.

Contractors will often charge a premium or cut corners in material or craftsmanship if you want to get a job done quickly.

If one wants a high-quality job, it will not be done on the cheap, nor will it be done quickly. Craftsmanship takes time, and skilled laborers earn high dollars for a reason.

Getting it done cheaply means using lower-quality materials and craftsmen who are just starting out or not up to snuff.

The three-legged stool is most stable when all the legs are similar in length. If one is too long or too short, the stool is unstable. There is an inherent give and take between the three. The trick is to strike the right balance.

Chapter Twenty-Two

Play it Again, Sam

THERE IS NEVER ENOUGH **time to do it right, but always enough time to do it over.**

Do not let the perceived pressure of time allow you to cut corners to meet a deadline. Cutting corners most likely will get caught by a vigilant quality control inspector. Making the deadline was a temporary victory. Now, one must remove the shoddy quality and do it right the a second, third, or fourth time. The additional work usually is not covered in contract costs, resulting in a loss of profit for the constructor.

Additionally, when workers touch that item, they are unnecessarily exposing themselves to additional hazards with the possibility of bodily harm to themselves or others. Slow down, do it right. Take the ass chewing for missing a deadline if you must. It will be less intense than explaining to your construction manager, project manager, or owner/client why you must do it a second, third, or fourth time.

FUN FACT: While the title of the chapter is often associated with the 1942 movie *Casablanca*, starring Humphrey Bogart, it is never said in the film. The actual line is "Play it, Sam."

Chapter Twenty-Three

Just Beat it

BEAT IT TO FIT, **paint it to match** When you desperately need to have it done now. When quality is not a concern. When the parts don't quite fit as designed, and it will take too long to modify them or get the appropriate materials, what is one to do?

Grab the universal wrench and crank it down. Grab the internationally known persuader (B.F.H.) and bash it together. Grab the heat torch and expand the metals so they can fit.

Once that is done, quality now matters – at least the appearance of quality. It is time to find the matching paint, covering, or cladding to cover up the rushed job.

This is not an approved procedure, but it is a BOLO - Be On the Look Out! Unscrupulous workers will take shortcuts.

Chapter Twenty-Four

I am all out of Names

"I'm here to kick ass and take names..and I'm all out of names"

WAKE UP. KICK-ASS. REPEAT
This is more of a motto or way of life than a colloquial saying. I attribute this to one of the contractor project managers building a job for us. He came on the site to lend leadership and knowledge to a site team on the verge of utter failure. He is a no-nonsense, get it right the first time, no excuses kind of guy, precisely what the contractor needed.

He was a joy to watch. He was holding court with his frontline leadership and supervisors, holding them accountable for their failures, praising them, and celebrating their successes. Those who could not cut the mustard or change to meet the new requirements were politely told to go elsewhere.

While he was no-nonsense, he was very personable and a pretty fun guy with a great sense of humor. The trick was always knowing when to reveal that part of himself and when to keep it under wraps. He walked that tightrope very well.

Because of his work ethic, leadership, and personal accountability, we finished the job on schedule. Something that in that region had never been accomplished before. Christmas came about a month before the completion of the job. I went out to purchase a coffee mug with the phrase printed. He smiled when he received it and appreciated the gesture. A couple of years later, he sent me a text to catch up on projects and life. In the text was a picture of his coffee mug, complete with the telltale signs of excessive coffee use.

Chapter Twenty-Five

Chicken Salad

CHICKEN SALAD FROM CHICKEN **shit**
A kin to the old "we will have to do more with less" adage, making chicken salad out of chicken shit, was used on a job site by a contractor's construction manager when we had the world stacked against us. The weather had turned ominous and flooded our job site. High winds damaged facilities and equipment. Morale was down. Worker Absenteeism was up. I often have said that construction is not a game for optimists. In construction, you plan for the worst, hope for the best, and are elated when it's something in the middle. But in this particular moment, there was no sense in being pessimistic. Pretty much every bad thing that could happen did. The CM stepped up, put his best foot

forward, and got to work. All that chicken shit on the job site was slowly transformed into chicken salad. But just like chicken salad, it wasn't the prettiest of things, but it was at least palatable.

Chapter Twenty-Six

Emojis don't count

LEAVE YOUR EMOTIONS AT **the gate.**
Construction is a people business. People come to work with various backgrounds, life goals, home problems, and work ethics. All of these are what make us human. It makes us an individual and not a robot. That said, on the job site, particularly in high-danger or high-risk jobs, those human qualities sometimes need to take a back seat to the cold and calculating part of the business.

Often what we must do and want to do will conflict with one another. If you are on the job, then focus should be the watchword. I often told my folks to leave their emotions at the gate. Not to dehumanize them, but more to get them focused. You cannot, or should not, oversee a complicated critical lift or execute the start-up of a plant if you are emotionally compromised or unfocused. Being clear-headed and of sound mind is essential. Knowing your people is also crucial. When Joe or Bob seems distracted, it is incumbent upon you as the leader to temporarily clarify the situation or remove them from the job. These are sound business decisions that may sound impersonal at the surface, but upon reflection, it shows the human side of people. It shows that you care enough to stop the work or remove distracted people from the job so they can go home in one piece.

Chapter Twenty-Seven

Do it for the Fishes

DON'T TELL ME WHAT I can't do; tell me how you can help me do what needs to be done.

How many times have you been trying to solve a problem, only to have the negative Nelly shoot them down one after another? The only contribution to the conversation is what cannot be done.

Safety folks tell you what you cannot do – OSHA/MSHA says you can't, they say. Quality assurance/control folks tell you what you cannot do – the spec says. Engineers tell you what you cannot do -it violates code X.Y.Z para-D. The environmental/regulatory folks tell you it violates the clean water, birdies, and fishes act of 19XX something or other.

With all the negativity, it is incredible that anything has ever or will ever be built. Please don't get me wrong. Safety, quality, environmental and regulatory, and design engineers have a crucial role in any project. I like clean water, cute birds, and fishes with the correct number of fins, but there ARE ways to solve problems. The people in those fields will tell you what you cannot do so that you don't get fined or go to jail, and they can say, "I told him not to do that."

The true professional in those fields will keep you out of jail and money in your pocket, but the most significant difference is they will tell you HOW you CAN accomplish the task or solve the problem. Some even become creative with interpreting regulations, codes, and statutes to support legal, lawful, and moral task accomplishment.

If you start your problem-solving sessions by acknowledging there are things we cannot do and focusing on what we can do, everyone wins- including the birds and fishes.

Chapter Twenty-Eight

What Time Is It?

I DON'T NEED TO **know how to build the watch, only what time it is!**

This is one of my all-time favorites. It was used by a good friend, fellow academy grad, and my deputy for construction on a job site up north. With similar backgrounds, I knew exactly where he was coming from. I need a simple yes or no, a simple the truck is late, or someone got hurt. But instead, the person being addressed feels the need to paint a holistic picture, complete with a timeline. There often is no time for the diatribe in matters requiring immediate decisions. Get to the point!

As a reminder, I used to keep a 3x5 index card on my desk with the letters B-L-U-F in a sizeable handwritten font. Bottom Line Up Front. Please get to the punchline; we have time for the after-action report later. Now time is of the essence. I would hold that 3x5 card up when someone was rambling so they could see it and understand the information I needed at the time was critical. I can always ask the who, the how, and the why later. It took a while, but eventually, folks that worked for me understood and would start their comments with "Hey Boss – BLUF – so and so <insert description here>" then proceed with the rest of the story once I acknowledged what they said or would answer my follow-on questions.

I wasn't trying to be an ass, although I have been accused many times. Instead, I needed to know whether this required upline reporting, action to be taken immediately, or if I had the luxury of listening to the verbal picture about to be painted.

Chapter Twenty-Nine

The Pen(cil) is Mightier than the Sword

There comes a time in every project when the engineer must put down the pencil so the work can begin.

Engineers are tinkerers by and large. Always on the quest to make something more efficient, smaller, powerful, or with a host of new features. This is not a swipe at engineers; I am one as well. It is just in our nature. But given their propensity to keep "improving" a design, someone must stop the madness so the work can begin. Said another way, don't let well perfect be the enemy of good enough. I was taught if it is safe, meets specifications, and works,s move on; you've done your job. Let the following group do theirs!

There is a saying in the military that the 80% plan violently executed was much better than the perfect plan never executed. That certainly sums up the thought here.

Chapter Thirty

Designed to a "T"

THE ENGINEER DESIGNS IT to the decimal point; the drafter draws it with a pencil, the fabricator marks it with chalk and the laborer cuts it with an ax.

Here is the difference between the academic world of engineering and the front line, where the men and women get the work done.

The precision with which something is designed and the realities of the end user are in stark contrast here. It also symbolizes the loss of fidelity as ideas are passed from engineer to fabricator to end user.

This is not to say that modern construction workers are Neanderthals and can't make this to specification. On the contrary, with the accuracy of laser survey, precision machining, and large-scale 3-D printing (as an aside, check out 3-D concrete buildings- we live in a fantastic age). The idea that achieving the design is not possible is quickly fading. It is an excellent reminder to the engineer to get out and see the field now and again to appreciate the limitations on the front line. The engineer is limited only by imagination; the construction worker is limited by many more elements – time, materials, weather, skills, etc.

Chapter Thirty-One

In the Navy

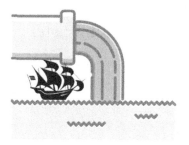

THE UNITED STATES NAVAL Academy Piping Specification:

Dout > Din

Seen on an engineer's whiteboard. While it is intuitively obvious to the casual observer, I will have to explain to the lovely graduates of the 4th best military academy in the United States:

1. West Point

2. Air Force

3. Coast Guard – yes, there is a US Coast Guard Academy at King's Point, NY,

4. Navy – Don't believe me? *Read **Discipline: The Annapolis Way: Lessons from the Nation's 4th best Military Academy**, an Amazon Best Seller by Mike Nemeth.*

You see, the pipe diameter of the outside of the pipe *MUST* be larger than the inside diameter. It is just that simple. However, some do need a little help. By the way, do not look for this specification in ASME B31.3. as the committee has yet to adopt it.

Go Army – Beat navy

P.S. - It is all in good fun. I love my fellow academy graduates from Canoe University, except when Army is playing football against them. I would not spend time making fun of them if I didn't care about them. Besides, in the end, we are all on the same team. Team America!

Afterword

I have done my best to cite these sayings as accurately as I can recall. I fully acknowledge others may have heard some of these sayings from other locations, vocations, or professions, but it was on the job site where I was exposed to these first, thus why I have included them here. If you want to argue they come from somewhere else, write your own book. ;)

Acknowledgement

A special thanks to the following individuals who have significantly impacted me as a construction professional and a person. I am grateful they took the time to coach, mentor, and teach me along the way.

Each is a true professional in their chosen discipline, be it safety, project management, or construction leadership. They could tactfully call me out when I wasn't thinking straight, teach me before things got out of hand, and when things got rough, always there for an after-work beer or "sweet tea" to help solve our problems and world peace. Not only were they workmates, but they also rose to the level of Friends.

Joe Allen	John Frederick	Mark Meade
Craig Boyd	Paul Furst	Kevin Miller
Wayne Chandler	Don Henley	Kristen Park
John "Big" Cummings	Charlie Holland	David Porsche
Greg Dingwall	Simon Home	Collin Pryor
Drew Edwards	Clint Lesyk	Jim Volker
Joseph Fess	Gabriel Marquez	Chris White
Scott Francis	Mike McDonald	

The following construction professionals impacted my professional growth as well and must be recognized though they have since passed. Be thou at Peace.

Don Ollar
Merle Powers
David Shimek
David Spear

Works Cited

- Nemeth, Mike. (2017). *Discipline: The Annapolis Way: Lessons from the Nation's 4th best Military Academy*

- R.S. Means Company. (2014). *RSMeans Heavy Construction Cost Data 2015*. Rsmeans.

- Sutherland, Jeffrey Victor(201). *Scrum : The Art of Doing Twice the Work in Half the Time*. London, Rh Business Books.

- Wikipedia Contributors. "Hurricane Ike." *Wikipedia*, Wikimedia Foundation, 18 Jan. 2020, en.wikipedia.org/wiki/Hurricane_Ike.

All Illustrations were created by the author with the use of Canva Pro at canva.com

Cover photo used with permission from Canva Pro.

About the Author

Sean was born in an Army Hospital in then-West Germany. He bounced around the globe from duty station to duty station with his parents and younger brother. When it was time to move out and be on his own, the Army was the only life he had known. Sean secured an appointment to the United States Military Academy at West Point, NY where he graduated with a Bachelor of Science in Civil Engineering. This was the start of his 25 years of service in the Army where he served in multiple positions and posts around the globe. While in uniform, The US Army Corps of Engineers offered him the opportunity to earn a Master of Science in Civil Engineering (Structural) from the University of Massachusetts – Amherst.

Serving in the U.S. Army Corps of Engineers, he was always around construction and project management activities.

After the Army, he moved on to become a construction manager in one of the largest oil and gas companies in the world. Working in many different nations, and climates, he found a similar brotherhood to the army. He has held jobs in the areas of project management for the last 2 years in commercial construction and most recently in renewable energy.

He resides in his home state of Texas with his wife and 4 kids. Oh, and a very loyal Husky who cannot stand to be away from him.

Made in the USA
Middletown, DE
06 September 2022